MW01258698

CROCODILES

AUSTRALIAN ANIMAL DISCOVERY LIBRARY

Lynn M. Stone

Rourke Corporation, Inc.
Vero Beach, Florida 32964

PHOTO CREDITS

All photos © Lynn M. Stone

ACKNOWLEDGEMENTS

The author thanks the following for photographic assistance: Australian Overseas Information Office, Northern Territory Protocol Office, Northern Territory Conservation Commission

LIBRARY OF CONGRESS

Library of Congress Cataloging-in-Publication Data

Stone, Lynn M.

Crocodiles / by Lynn M. Stone.

p. cm. — (Australian animal discovery library)

Summary: Discusses the physical characteristics and behavior of the crocodile, with an emphasis on its presence in Australia.

ISBN 0-86593-060-0

1. Crocodiles—Juvenile literature. [1. Crocodiles—Australia] I. Title. II. Series: Stone, Lynn M. Australian animal discovery library.

QL666.C925S78 1990

597.98—dc20 90-31575

CIP

AC

Printed in the USA

Saltwater Crocodile

TABLE OF CONTENTS

CROCODILES

Do crocodiles smile? No, not even when they are well fed.

They just seem to smile because they have long jaws with crooked edges and many flashing teeth.

Those teeth—and powerful jaws—help make Australia's saltwater crocodile *(Crocodylus porosus)* one of the world's most dangerous animals.

The saltwater croc is also one of the world's largest animals. A big male can weigh over a ton—2000 pounds!

Among the reptiles—snakes, lizards, turtles, alligators, and crocs—none is larger than the saltwater crocodile.

Saltwater Crocodile

THE CROCODILE'S COUSINS

The saltwater crocodile is one of 23 kinds, or **species,** of **crocodilians.**

"Crocodilian" is the term scientists have for alligators, crocodiles, and gharials.

The crocodilians look and act very much alike. Crocs, however, have narrower jaws than alligators. Gharials have narrower jaws than crocs.

There are other differences in the crocodilians which you cannot see.

The saltwater croc's closest cousins are the other 12 crocodiles. One of them, the small freshwater crocodile *(Crocodylus johnstoni),* lives near the saltwater crocs in Australia.

American Alligator

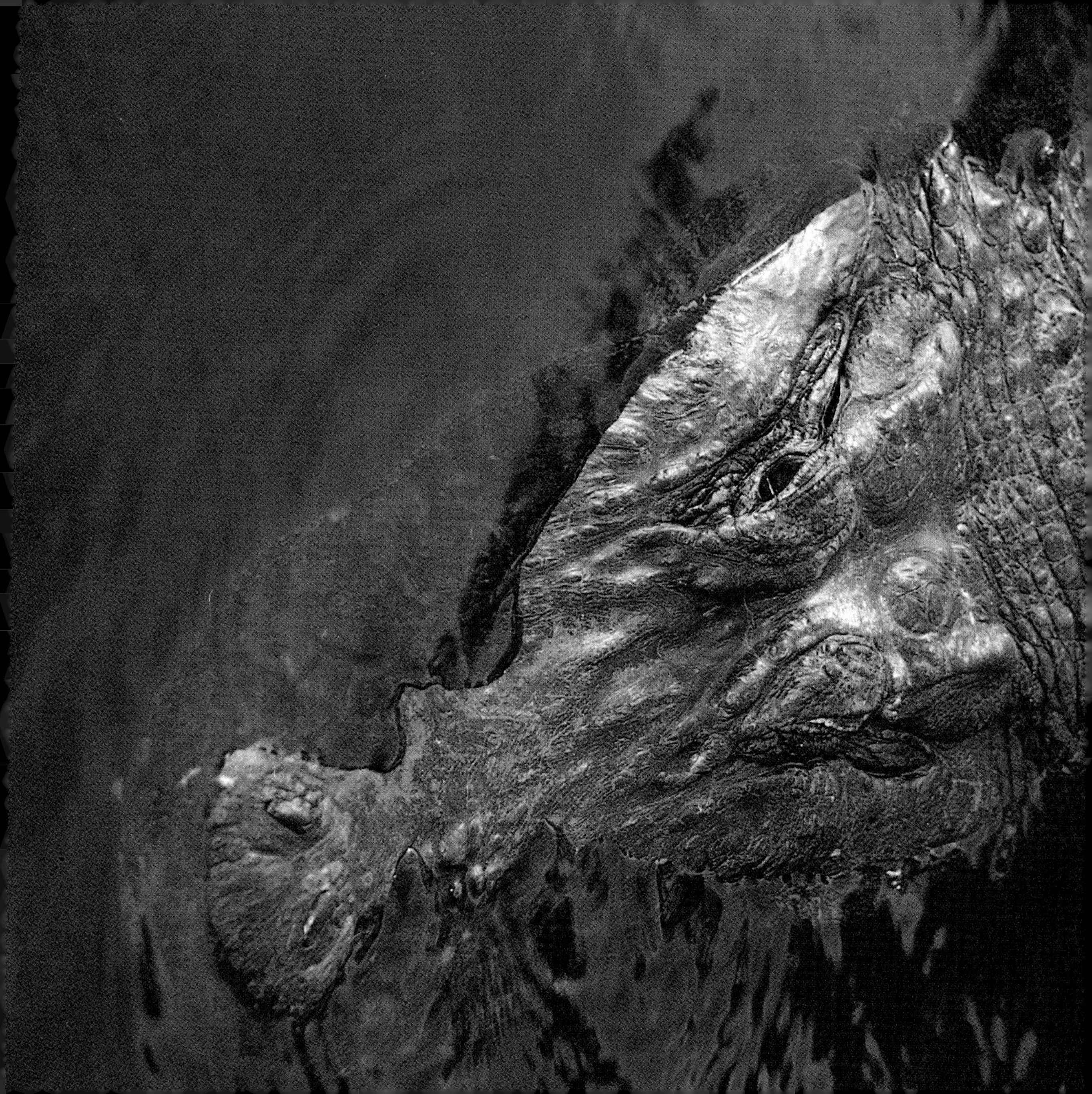

HOW THEY LOOK

Picture the crocodile as a giant, water-loving lizard. It has long, flat jaws and stubby legs with webbed hind feet.

The croc's body and tail are long, like a streamlined log. The tail is used for swimming power.

The croc is covered by leathery scales. Along its back and tail are ridges—tiny hills and flaps of skin.

Saltwater crocs can be huge. Males reach 15 feet in length. There are stories of saltwater crocs more than 20 feet long.

WHERE THEY LIVE

Australia's saltwater crocs live only near the sea coasts of north and northeastern Australia. Most of the saltwater crocs are in the Northern Territory. Others live in the states of Western Australia and Queensland.

Saltwater crocs also live north of Australia and in such Asian countries as Sri Lanka, India, and Burma.

Saltwater crocs are not always found in saltwater. They live in rivers, swamps, and ponds, too.

Crocodile Country,
Northern Territory

Saltwater Crocodile

HOW THEY LIVE

Saltwater crocodiles spend most of their time in the water. They are made for life in the water. Flaps of skin keep water out of their throat, eyes, and ears when they swim. The shape of their bodies helps them swim easily.

Crocs usually leave the water only to nest and sun themselves. Sunshine helps keep crocs warm on cool days.

If a croc becomes too warm, it returns to the water or finds a shady river bank.

CROCODILE BABIES

Baby crocodiles are hatched from eggs. A mother saltwater croc lays about 55 eggs in her nest.

The saltwater croc's nest is a mound of mud and plants. The mother crocodile usually stays close to her nest. Often she will attack anything or anyone who comes near.

When they hatch, the babies are helped from the nest by the mother.

Saltwater crocs become adults at about 14 years of age. Some may survive to reach 90 or 100 years.

Saltwater Crocodile with Bird

PREDATOR AND PREY

The saltwater crocodile is a **predator** because it feeds on animals. The animals which it eats are its **prey.**

The saltwater crock eats turtles, pigs, mudcrabs, water birds, **wallabies,** and other animals.

A croc can swim with only its eyes and nostrils above water. When it spies prey, it glides underwater toward the prey. The croc grabs the victim in its jaws.

A croc can lunge over one-half of its body length. By lunging, it can catch animals on the shore and in low branches over the river.

Croc Prey: Jaribu Stork, N. Territory

CROCODILES AND PEOPLE

Most Australians are proud of their huge crocodiles. At the same time, they treat the crocs with fear and respect.

People in Australia are warned not to swim or take small boats into crocodile country. People who are careless are sometimes attacked and killed by saltwater crocs.

Crocodiles are important in the paintings and religion of the Australian **Aborigines.** Aborigine tribes were the first settlers of Australia.

THE FUTURE OF THE CROCODILE

Between 1945 and 1972, the saltwater crocodile was hunted for its meat and skin. The crocs began to disappear.

Since 1972, the croc has been protected from hunting. Only the aborigines still take crocodiles and their eggs.

Now, the saltwater croc is increasing. Its **habitat,** the wetlands where it lives, is healthy.

To keep a supply of crocodile meat and leather, Australians have begun crocodile farms.

The future is bright for Australia's wild saltwater crocs. But even such good news won't make a crocodile smile.

Glossary

Aborigine (ah bore IDJ in ee)—the original or native people of a place

crocodilian (crock a DILL ee un)—the alligators, crocodiles, and gharials

habitat (HAB a tat)—the kind of place an animal lives in, such as a river

predator (PRED a tor)—an animal that kills other animals for food

prey (PRAY)—an animal that is hunted by another for food

species (SPEE sheez)—within a group of closely related animals, one certain kind

wallaby (WALL a bee)—a type of kangaroo found in Australia

INDEX